Write It Right

Writing a Lab Report

By Cecilia Minden

Published in the United States of America by
CHERRY LAKE PRESS
Ann Arbor, Michigan, 2395 South Huron Parkway, Suite 200
www.cherrylakepublishing.com

Reading Adviser: Marla Conn, MS, Ed., Literacy specialist, Read-Ability, Inc.
Book Designer: Felicia Macheske
Character Illustrator: Carol Herring

Photo Credits: © wavebreakmedia/Shutterstock.com, 13; © Narongrit Pootson/Shutterstock.com, 17

Graphics Throughout: © simple surface/Shutterstock.com; © Mix3r/Shutterstock.com; © Artefficient/Shutterstock.com; © lemony/Shutterstock.com; © Svetolk/Shutterstock.com; © EV-DA/Shutterstock.com; © briddy/Shutterstock.com; © IreneArt/Shutterstock.com

Library of Congress Cataloging-in-Publication Data

Names: Minden, Cecilia, author. | Herring, Carol, illustrator.
Title: Writing a lab report / by Cecilia Minden ; illustrated Carol Herring.
Description: Ann Arbor : Cherry Lake Publishing, 2020. | Series: Write it right | Includes bibliographical references and index. | Audience: Grades 2-3
Identifiers: LCCN 2019033439 (print) | LCCN 2019033440 (ebook) | ISBN 9781534159037 (hardcover) | ISBN 9781534161337 (paperback) | ISBN 9781534160187 (pdf) | ISBN 9781534162488 (ebook)
Subjects: LCSH: Technical writing–Juvenile literature. | Report writing–Juvenile literature.
Classification: LCC T11 .M534 2020 (print) | LCC T11 (ebook) | DDC 808.06/66–dc23
LC record available at https://lccn.loc.gov/2019033439
LC ebook record available at https://lccn.loc.gov/2019033440

Cherry Lake Publishing would like to acknowledge the work of the Partnership for 21st Century Learning, a Network of Battelle for Kids. Please visit *www.battelleforkids.org/networks/p21* for more information.

Printed in the United States of America
Corporate Graphics

CHERRY LAKE PRESS

Table of
CONTENTS

CHAPTER ONE

You Can Be a Scientist!

Studying science is one of the best ways to learn about the world around you. Experimenting gives you a chance to figure things out through exciting hands-on exploration.

There's nothing like the thrill of conducting your own experiments. You can do them in a lab at school. You can even do experiments at home!

What science experiments have you tried?

Scientists keep track of their experiments by writing a **lab report**. A lab report describes the findings of their research. It explains what they've discovered. It also details the conclusions they made based on those findings. There is a method all scientists follow to conduct scientific experiments. It is called the **scientific method**.

CHAPTER TWO

The Scientific Method

The scientific method is a way to ask and answer questions. Scientists research and make **observations**. They also do experiments. There are generally five steps in the scientific method:

1. Make an observation. What things make you curious to learn more? You may have observed that clouds are different shapes. Do you want to know why?

2. Ask a question. This narrows the focus of your observation. Why are clouds different shapes?

3. Make a **hypothesis**. A hypothesis is your best guess for an answer to your question. It is based on what you already know. You know clouds are tiny drops of water pushed upward by the air current. Your hypothesis could be that the speed of the air current causes the shape of the cloud.

4. Run an experiment. Design an experiment that will test your hypothesis. Research and monitor cloud formations and air currents.

5. **Analyze** your findings and draw a conclusion. Study the results. Do they support your hypothesis?

Use the information you know to think of a hypothesis.

ACTIVITY

Make a List

In this activity, you will brainstorm hypotheses.

INSTRUCTIONS:

1. Write down at least three hypotheses you would like to test.
2. Think about what you already know.
3. Think about what you want to learn.
4. Be creative!

Sample Activity

- You will do better on a test if you get a good night's sleep.
- Drinking many high-sugar sodas is not healthy.
- Dark colors absorb more sunlight. ✓

The Scientific Method

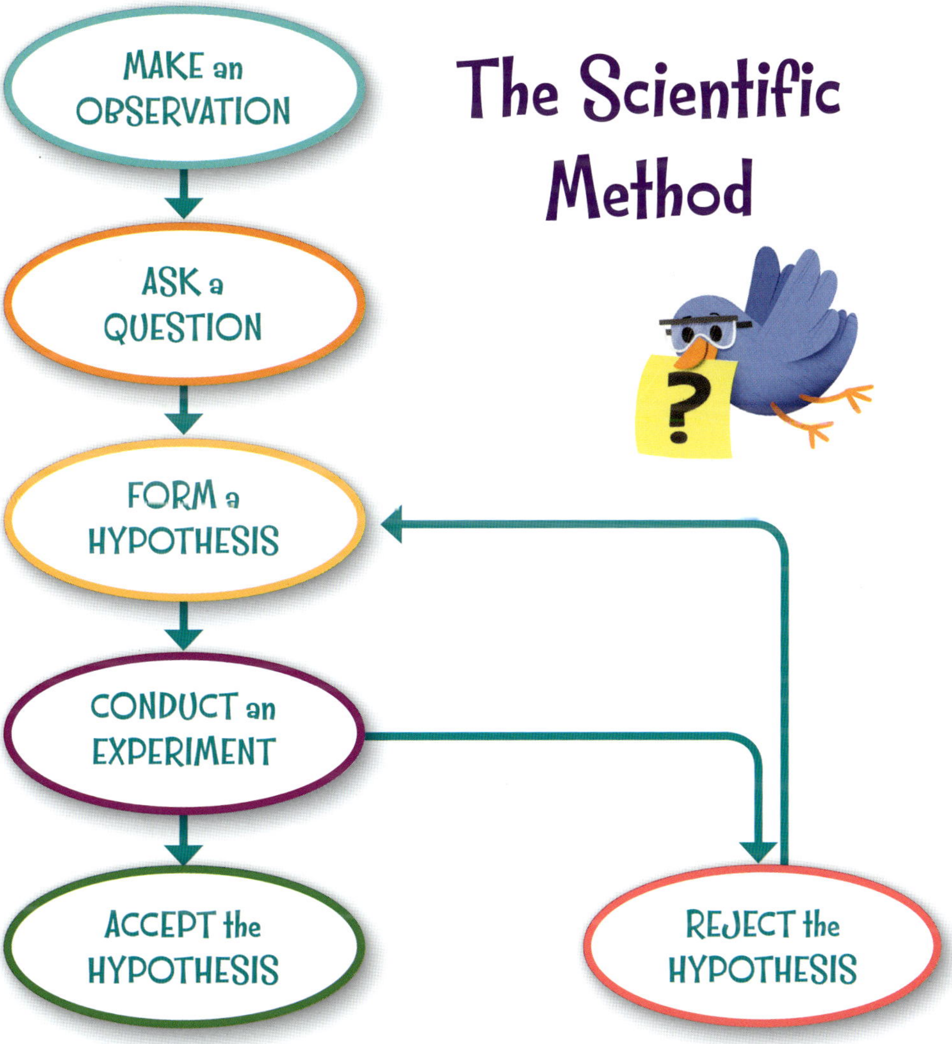

CHAPTER THREE

Let's Experiment!

Begin your report by explaining the purpose for your experiment. Write the question you want to **investigate** in this section. You can also explain why you want to conduct the experiment. This can be something you observed. It could also be something you read about.

State your hypothesis in the next section. Remember that a hypothesis is a guess. It predicts what you expect to happen. You should always include your original hypothesis in your lab report. Include it even if it is proven wrong by your experiment. An incorrect hypothesis can help provide you with new questions. You can use them in another experiment.

ACTIVITY

The First Steps!

Your lab report should begin with explaining what you are studying.

HERE'S WHAT YOU'LL NEED:

- Notebook paper
- Pen
- Computer (optional)

INSTRUCTIONS:

1. Write your name and the date at the top of the paper.
2. Write the project title of the experiment.
3. Write the question you want to answer.
4. Fill in your hypothesis.

Sample Activity: Lab Report

Name: Taylor **Date:** September 19, 2020

Project Title: Testing Which Colors Cause Ice to Melt Faster

Question (purpose or reason for my experiment): I observed that direct sunlight feels different when I am wearing different colors of clothing. The question I want to answer is: Which colors keep me cooler? I will use ice to determine this hypothesis.

Hypothesis: Dark colors absorb more sunlight; therefore, ice on dark colors will melt faster.

CHAPTER FOUR

What Is Your Experiment?

The next step is to describe your experiment. This information explains the process you used. It includes the materials you used. It also includes the steps you took to carry out your experiment.

List all of your materials. Record how much of them you used. Be specific. Describe the amount in weight, length, or any other measurement you need. Include the brand name of the materials as well.

Now describe the steps you used to conduct the experiment. They should be listed in the exact order you did them. Your instructions should be clear enough to help others conduct your experiment just as you did. Include many details. Explain how long the experiment took. This section should also detail preparation work, such as labeling your materials.

Type or write down a complete list of materials.

ACTIVITY

Keep at It!

Continue to write your lab report.

INSTRUCTIONS:

1. Add the list of materials you used.
2. Include the steps you used to conduct your experiment.

Materials:

- 4 different colored square pieces of paper larger than an ice cube
- 1 plastic ice cube tray holding 4 equal size cubes
- Small clear bowl
- Tray to carry items outside
- Stopwatch

Methods:

1. I cut my colored paper into four even squares. I used black, blue, red, and white.
2. I poured water into a plastic ice cube tray. I placed the tray in the freezer for the ice to harden.
3. I placed the four paper squares and a clear bowl on a tray.
4. Once the ice hardened, I placed four ice cubes in the bowl and took the tray outside.
5. I kept the bowl in the shade. I placed the four pieces of paper on a sunny sidewalk.
6. I placed one ice cube on each piece of paper and started my stopwatch.
7. I recorded the moment each ice cube completely melted on the paper.

CHAPTER FIVE

What Did You Learn?

The next section of your lab report contains your results. This is where you include the **data** or measurements you recorded. You can include photographs or charts to help explain the data. This section presents your results directly. It does not discuss or explain your thoughts on the experiment's outcome.

The conclusions section is the final part of your lab report. This is where you compare your hypothesis to the results of your experiment. Was your hypothesis correct? Use your data and observations to explain your findings.

You may also wish to discuss the reasons your results turned out the way they did. It is possible that your methods affected the results.

Keeping records is an important part of scientific experiments.

ACTIVITY

Review Results!

Add your results and conclusions to your lab report.

INSTRUCTIONS:

1. Organize the results of your lab report.
2. Add your conclusions.

Sample Activity: Lab Report (Continued)

Results (the findings of my experiment):

1. The ice melted fastest on the black piece of paper.
2. The ice melted at the same rate on the red and blue pieces of paper.
3. The ice melted slowest on the white piece of paper.

Conclusions:

1. The ice melted fastest on the black paper because black absorbs the light.
2. The ice melted slowest on the white paper because white reflects the light.

CHAPTER SIX

Completing Your Report

Bar graphs are one of the most common types of graphs used in lab reports. A bar graph is good for comparing things.

ACTIVITY

Making a Graph

Add your results and conclusion to your lab report.

HERE'S WHAT YOU'LL NEED:

- Graph paper
- Pencil
- Ruler

INSTRUCTIONS:

1. Create a bar graph using the results of your experiment.
2. Label the x-axis with the colors of paper.
3. Label the y-axis with the times it took the ice to melt.
4. Draw the height of the bar at each color to match the time it took the ice to melt.
5. Color in each of the four bars in your graph.

Sample Bar Graph

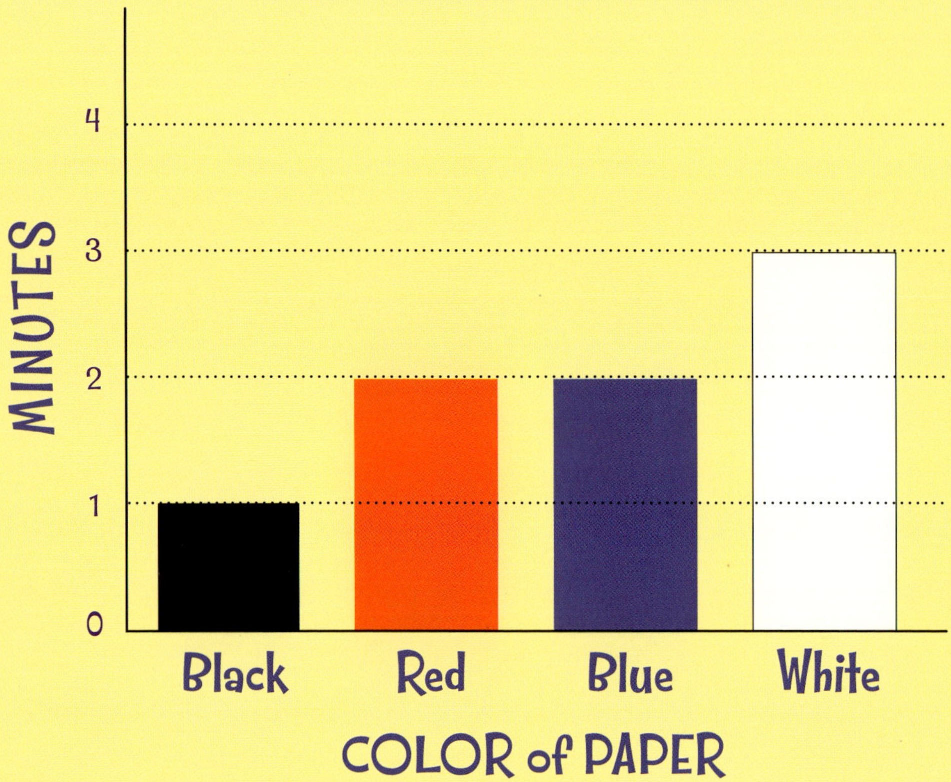

What will your next experiment be?

ACTIVITY

Almost Done!

Before you hand in your lab report to your teacher, check everything one more time.

- Did I write my name and the date on my lab report?
- Did I write the name of my project title on my lab report?
- Did I write down my question and hypothesis?
- Did I describe my experiment carefully?
- Did I list all the materials and steps I used? Did I record my data accurately?
- Are the graphs and charts I created easy to understand? Are they properly labeled?
- Did I clearly state my conclusions?
- Is everything spelled correctly in my lab report?

GLOSSARY

analyze (AN-uh-lize) to study something carefully in order to understand it

bar graphs (BAHR GRAFS) charts that compare information by showing rectangular bars of different lengths

data (DAY-tuh) information that can be used to plan or make conclusions

hypothesis (hye-PAH-thih-sis) an idea that could explain how something works and is tested in an experiment

investigate (in-VES-tuh-gate) to gather information about something

lab report (LAB rih-PORT) a written document that describes and analyzes an experiment that explores a scientific concept

observations (ahb-zur-VAY-shuhnz) things you have noticed by watching carefully

scientific method (sye-uhn-TIF-ik METH-uhd) a way to ask and answer scientific questions by making observations and doing experiments

For More
INFORMATION

BOOKS

Chatterton, Crystal. *Awesome Science Experiments for Kids: 100+ Fun STEAM Projects and Why They Work!* Berkeley, CA: Rockridge Press, 2018.

Heinecke, Liz Lee. *Kitchen Science Lab for Kids: 52 Family-Friendly Experiments from Around the House.* Beverly, MA: Quarry Books, 2014.

WEBSITES

Lab Report Template
www.biologycorner.com/worksheets/labreport.html
Check out some helpful reminders on how to organize your lab reports.

Science Buddies—Steps of the Scientific Method
https://www.sciencebuddies.org/science-fair-projects/science-fair/steps-of-the-scientific-method
Explore the steps of the scientific method.

INDEX

About the AUTHOR

Cecilia Minden is the former director of the Language and Literacy Program at Harvard Graduate School of Education. She earned her doctorate from the University of Virginia. While at Harvard, Dr. Minden also taught several writing courses. Her research focused on early literacy skills and developing phonics curriculums. She is now a literacy consultant and the author of over 200 books for children. Dr. Minden lives with her family in McKinney, Texas. She enjoys helping students become interested in reading and writing.